This Book Belongs To

_____

_____

For my perfection of a daughter who never ceases to amaze me. You make me strive to be the best father possible.

Thank you Liz for your love and endless dedication on our amazing journey.

With enormous gratitude to all the women who make the decision to help families grow in one of the most meaningful ways possible.

And special thanks to the one woman who stepped forward to help our family become whole.

Daddy, What is an Embryo?
A Tale of Egg Donation

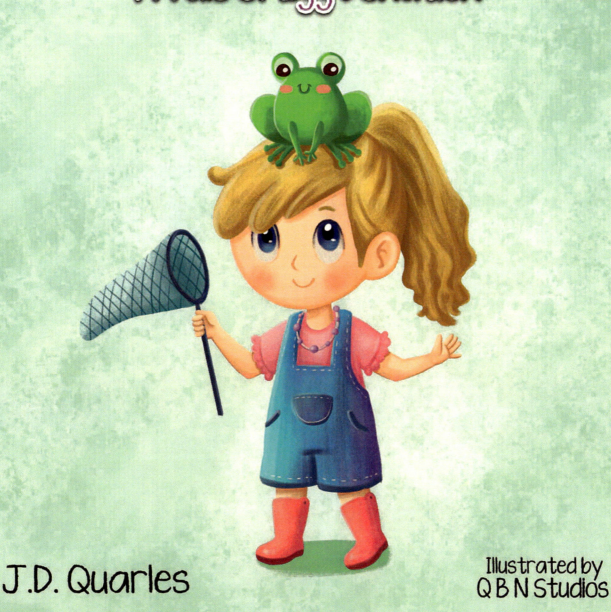

J.D. Quarles

Illustrated by
Q B N Studios

It was a beautiful morning as Ruthie and Daddy crept along the edge of the pond searching for the next bit of excitement.

A small splash caught Ruthie's eye and she quickly plunged her net into the water's edge.

"Did you catch anything?" asked Daddy.

"Just a bunch of gooey bubbles," said Ruthie.

"Those aren't bubbles," Daddy said, pulling a magnifying glass from his pocket. "Those are frog eggs! That tiny black dot inside each egg is a frog embryo."

"Daddy, what is an embryo?" asked Ruthie.

"Every embryo is a combination of tiny parts
(called cells) from the mommy and daddy.
Each of these frog embryos is the beginning of
what will grow into a baby frog," Daddy explained.

"Every animal starts out as an embryo; frogs, fish, birds, squirrels. Even you and I started as embryos," Daddy continued.

"Even me?" Ruthie asked in amazement.

"Even you," said Daddy. "You were smaller than this tiny frog egg when you started out."

"Did I start out as an egg too?" asked Ruthie.

"Yes! You started out as a human egg," Daddy said. "Actually, the egg was the beginning of your story."

"In fact, your mommy and I had to go to a special doctor to make the embryo that would become you," Daddy said.

"The doctor; were you sick?" asked Ruthie.

"No, not sick. It was because your mommy didn't have any usable eggs of her own," Daddy continued.

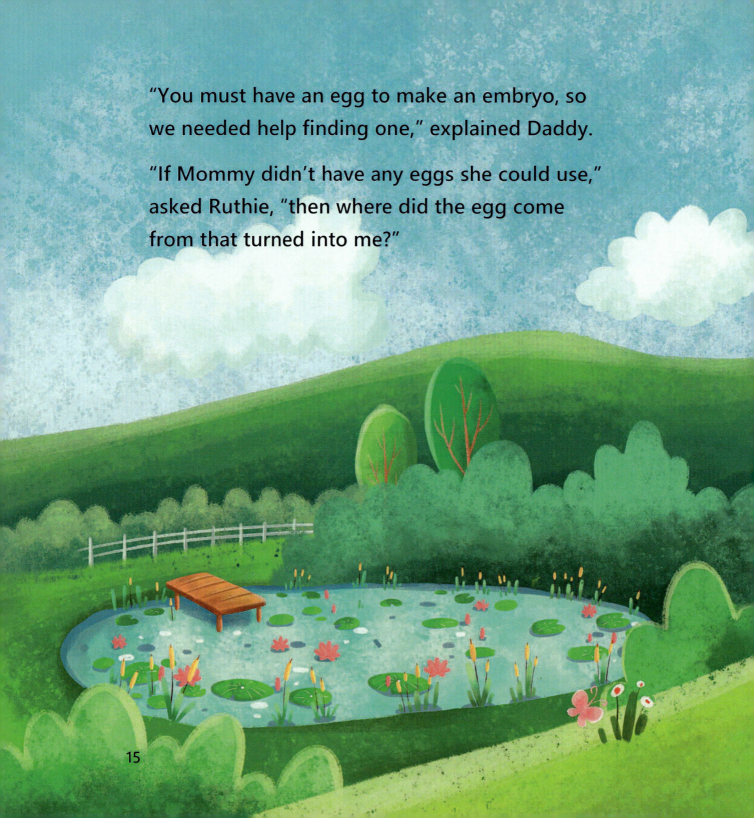

"You must have an egg to make an embryo, so we needed help finding one," explained Daddy.

"If Mommy didn't have any eggs she could use," asked Ruthie, "then where did the egg come from that turned into me?"

15

"The doctor had a list of many generous women, called egg donors. These women wanted to help families like your mommy and I have our own children by donating some of their extra eggs for others to use."

19

"We read through many profiles describing each egg donor; what they looked like, what their job was, what they enjoyed doing.

After reading about all the women, we found one that we knew was perfect for us."

"The doctor called the egg donor and told her that a family who couldn't have children on their own needed her help."

"Then, the egg donor went to the doctor's office and donated one of her eggs for us to have."

"In the laboratory, the doctor combined the donor egg with special cells from Daddy which caused the egg to become an embryo."

"Using a microscope, the doctor watched our embryo grow in the laboratory for a few days to make sure it was healthy."

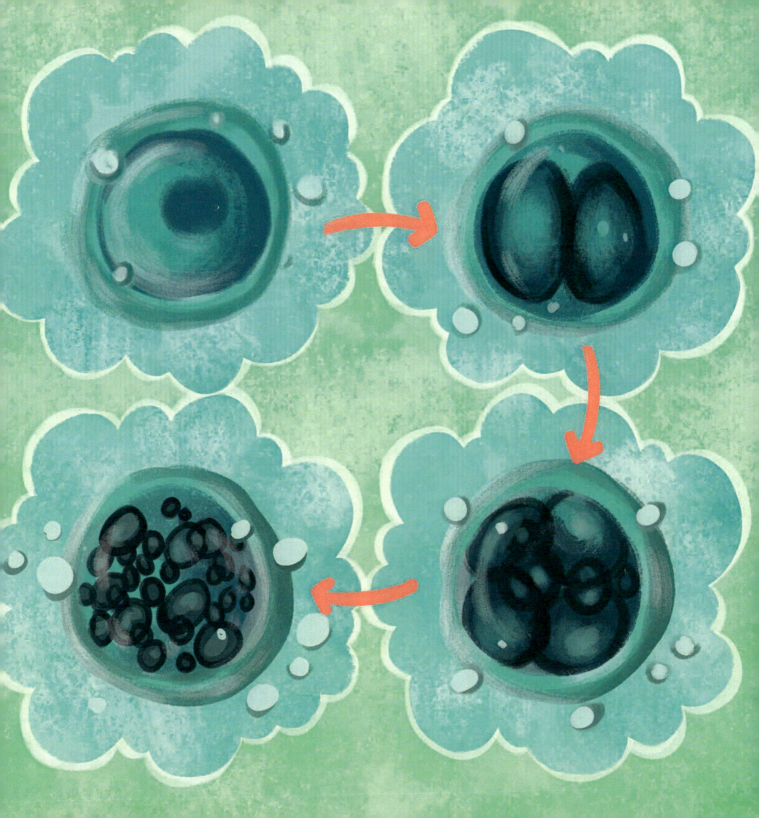

"Next, your mommy and I went to the doctor's office where the doctor put our embryo into your mommy's belly so that she would become pregnant."

"Our embryo grew and grew and grew inside your mommy's belly and turned into you!"

"After nine months, you were ready to be born. We went to the hospital where your mommy gave birth to you. We were so happy!"

"The egg donor made it possible for your mommy and me to have a child of our own. Even though we don't know the donor, she will always have a special place in our hearts for the amazing gift she gave us."

37

"Wow!" exclaimed Ruthie. "You and Mommy went through all that just to have me?"

"We sure did," Daddy said proudly. "It was the best decision we ever made."

"The beginning ..."

About The Author

After having his daughter at 46 years old,
using both a surrogate and egg donor,
J.D. Quarles wrote books in an effort to help
his daughter understand the amazing process
that brought her into this world.

While watching his daughter's comprehension
of this process develop, he decided to publish
the books to help others in similar situations.

Other Works By J.D. Quarles

Mommy, Who Is Miss Becky? A Tale of Surrogacy

Mommy, Who Is Miss Amy? A Tale of Surrogacy

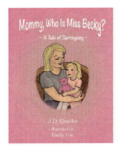 

Available through JDQuarles.com

Made in United States
North Haven, CT
26 February 2025

66291500R00029